laserpronet.com
Empowering the Laser Workforce

Name: _____

Cell/Text #: _____

What we do at LaserPronet

- Publication of Laser Tech Books
- Professional Development Services
    - Screening Exams
    - Professional Growth Plans
    - Professional Courses and Certifications

orders@laserpronet.com

Laser Procedures Essentials

Copyright © 2020

Sukuta Technologies, LLC

All rights reserved

Table of Contents

Topic	Page
1. Laser Procedure Content Outline	4
2. Laser Procedure Examples	8
3. Laser Resonant Cavity Fabrication/Alignment Procedures	15
4. Format for Standard Operating Procedures (SOPs)	19
5. Procedure Test Drive	24
6. Laser Tech Book Catalog, Courses and Certifications	25

1. Laser Procedure Content Outline

1.0 Cover Page

A Cover page must have at least fields for

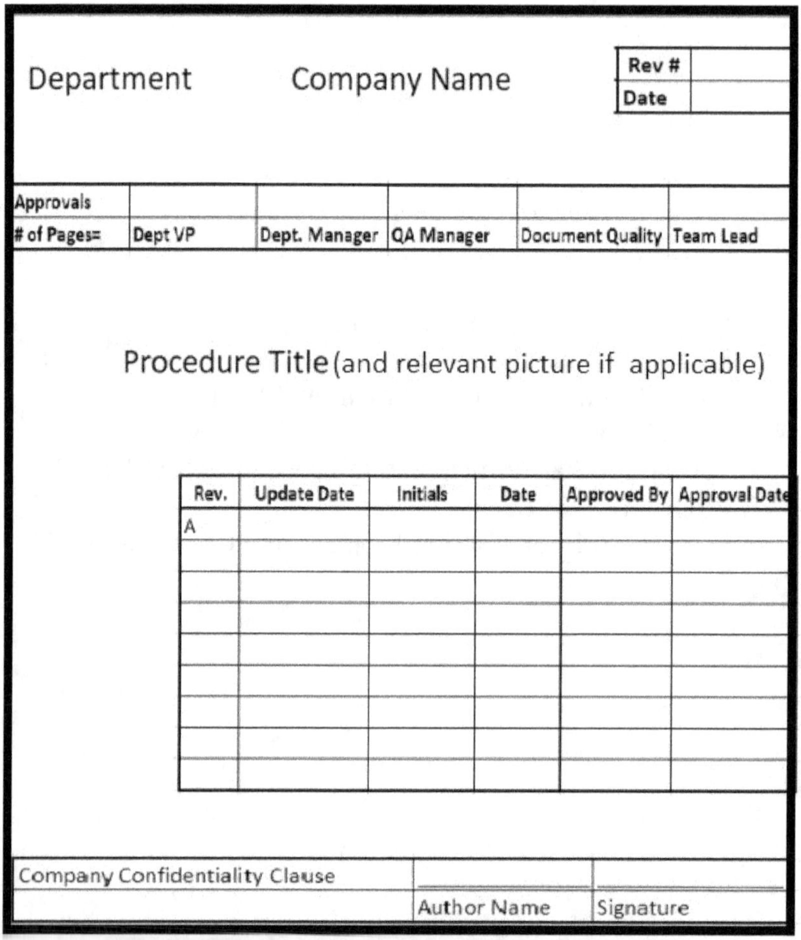

Department Company Name

Rev #	
Date	

Approvals					
# of Pages=	Dept VP	Dept. Manager	QA Manager	Document Quality	Team Lead

Procedure Title (and relevant picture if applicable)

Rev.	Update Date	Initials	Date	Approved By	Approval Date
A					

Company Confidentiality Clause		
	Author Name	Signature

Figure 1.1. Illustration of a blank procedure cover page.

a. Company name, Department, Procedure Title and Number/ID
b. Author(s) Name(s), Signature(s) and Date
c. Procedure Revision/Quality Control Section
    i. Revision number, date, description and who approved and their signature(s).

Table of Contents

Topic	Page

1. Preliminaries
   a. Objective/Purpose
   b. Scope
   c. Applicable Reference Documentation and Specifications
   d. Responsibilities
   e. Definitions/Acronyms
   f. Equipment Needed
   g. Warnings
2. Procedure
3. Sample Output, if applicable

1.1. Preliminaries

a. Objective/Purpose:
   i. *State why you are writing the procedure.*
b. Scope:
   i. *State what/where the procedure will be used.*
c. Applicable Reference Documentation and Specifications
   i. *State all documentation, and any resources, that were consulted for compiling the procedure, including standards and regulatory requirements.*
d. Responsibilities:
   i. *State who, or which department, is responsible in making sure that the provisions of the procedure are followed*
e. Definitions/Acronyms etc.
   i. *Define all highly technical terms/words/jargon used in the document that may be unusual to others* who may use the procedure.
f. Equipment Needed:
   i. *Numerically list of materials, tools, equipment, accessories, etc... needed to accomplish task(s)*
g. Warnings:
   i. *Alert user of any potential hazards while working on task(s) associated with this procedure.*

1.2. Procedure

h. Give step-by-step instructions on how to complete a task/goal
i. Start with safety steps/precautions before any work is done.
j. Include measures or rubrics that confirm procedure step completion.
   i. Include decision factors, such as what to do if a step-completion measure or rubric is not met.
k. Diagrams/Pictures/Drawings/Videos at each step are almost a must so that users can complete tasks "at a glance" in lieu of reading through each sentence or every step in the procedure all the time. Text-only procedures could be cumbersome, boring and even slow-down workflow.

1.3. Sample Output
if applicable
   Data in Tabular form
   Computation and Result(s) of Key Performance Indicators (KPIs)

2. Laser Procedure Examples

2.0 Cover Page *Example*

Laser Final Test			BMx			DATE: May 3rd, 20XX REV: #1	
APPROVALS:							
Total number of pages is 5	Vice President Engineering	Engineering Mgr	Manufacturing Mgr	Quality Assurance Mgr	Document Controller	Lead	

Getting Started with the BMx Spectrum Analyzer

REV	UPDATED BY	DATE	APPVD BY	DATE
A	Initial Release	JW		

Information contained in this document is considered to be confidential and is not to be used in any manner without the express permission of SJCC Laser Tech Department.

Signature and Date

Figure 2.1. Illustration of a generic test procedure cover page.

2.1. Preliminaries: *Examples*

a. **Objective/Purpose:** *Example*:

> *The purpose of this procedure is to explain the proper operation of a solid-state laser.*

b. **Procedure Scope**: *Example*

> *The purpose of this procedure is to explain the proper operation of the Model XXX 532nm Q-Switched Laser.*

c. **Applicable Reference Documentation and Specifications:** *Example*

> - *Device(s) User's Manual*
> - *ANSI/OSHA Laser Eye Safety Guidelines*
> - *Laser Operation SOP*
> - *Optics Handling, Inspection and Cleaning SOP*
> - *ESD Avoidance and Control SOP*

d. **Responsibilities:** *Example*

> *It is the responsibly of the XXX Department manager to verify that the provisions of this document are followed.*

e. **Definitions/Acronyms/Glossary:** *Example*
   i. *Define all highly technical terms/words used in the document that may be unusual to others who may use the procedure.*

1. Nd:YVO4: Neodymium doped, Yittrium Vanadate crystal
2. A laser pump transfers energy to the active medium.
3. A laser active medium coverts pump energy into laser radiation
4. The Fundamental wavelength of a laser is the wavelength that is produced by an active gain medium.
5. Etc..

f. **Equipment Needed:** *Example*
   i. *List of all materials, tools, equipment etc. needed to accomplish task(s)*

Materials, Tooling and Equipment Lists

Table 2.1 Materials List

Item	Qty	Description	Part No.	Manufacturer
1	Cut to Desired Length	Shielded 3 Conductor Cable, 22 AWG 0.118" OD		
2	2	Solder Sleeve Shield Terminator	S02-08-R	Raychem
3	4	Mini Sure-Seal Pins	330-8672-100	PEI-Genesis
4	4	Mini Sure-Seal Sockets	031-8703-100	PEI-Genesis
5	2 x 1"	Black 1.4" Dia Flexible Heat Shrinking Tube	PHS-016	SPC Technology
6	2 x 2"	Black 1/2" Dia Flexible Heat Shrinking Tube	PHS-032	SPC Technology
7	2 x 2"	Clear 1/4" Dia Flexible Heat Shrinking Tube	ES1000-NO.1-C1-X-STK	Raychem
8	1	Mini Sure-Seal Plug (4 conductors)	120-8552-102	PEI-Genesis
9	1	Mini Sure-Seal Receptacle (4 conductors)	120-8551-102	PEI-Genesis

Table 2.2 Tools List

Item	Description	Part No.	Manufacturer	Image
10	Standard Cutting Pliers	---	---	
11	Standard Utility Knife	---	---	
12	Standard Heat Gun	---	---	
13	Stripping Tool	3757-2	C.K.	
14	Crimping Tool	696202-1	Tyco Electronics	
15	Insertion Tool	MSS-2000	PEI-Genesis	

Equipment Needed: *Example*

1. *X-Head Q-Switched Laser*
2. *Control Module*
3. *Model XX Power Supply*
4. *Chiller*
5. *yyy nm laser safety eyewear*
6. *Power meter*
7. *Power detector*

    g. **Warnings**: *Example*

E1. Laser will shut down if operated past the maximum pump current.
E2. Ocular damage is likely if laser is operated without appropriate eyewear,

2.2. Procedure, *Examples*

Diagrams/Pictures/Drawings/Videos at each step are almost a must so that users can complete tasks "at a glance" in lieu of reading through each sentence or every step in the procedure all the time. Text-only procedures could be cumbersome, boring, and even slow-down workflow.

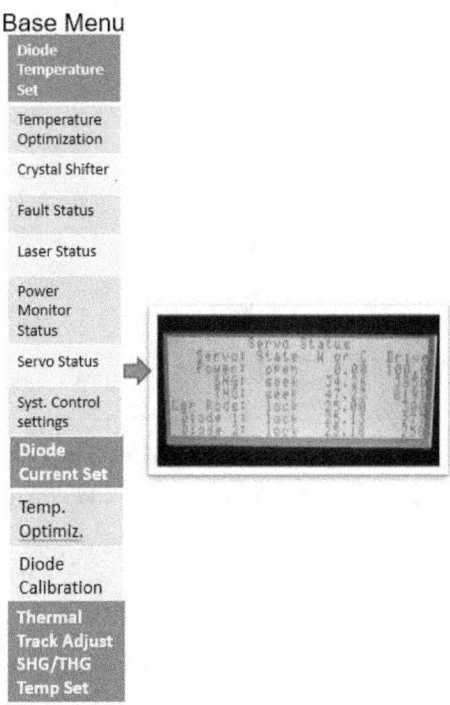

Figure 2.2 Example of how pictures and diagrams can add more clarity to a procedure step.

2.3 Sample Output, *Examples*

How sample characteristic output and/or KPIs guide procedure users to quickly detect faults if they differ from the expected

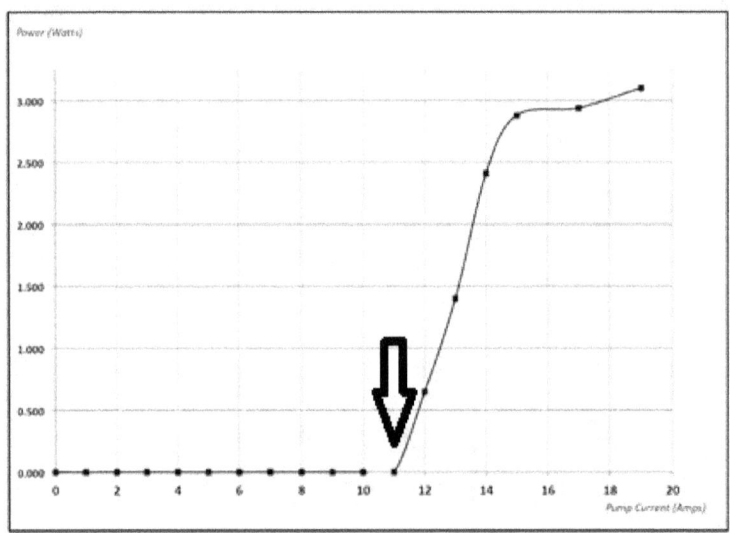

Figure 2.3 Graph quickly shows user how to quickly identify a laser's threshold pump input with minimal verbiage.

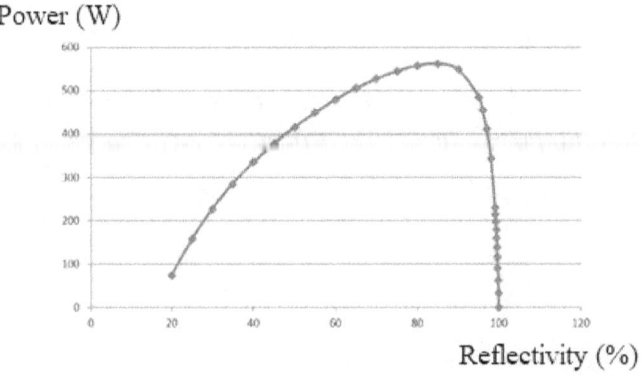

Figure 2.4 Expected laser power output pattern as a function of OC Reflectivity.

3. Laser Resonant Cavity Fabrication/Alignment

3.0 Cover Page – See Section 2.0
3.1 Preliminaries – See Sections 1.2 and 2.1
3.2 Laser Overview
　　List and overview how each laser component work. Typical components include
3.2.1　Pump
3.2.2　Active medium
3.2.3　Mirrors i.e. HR and OC
3.2.4　Q-switch, if applicable
3.2.5　Harmonics generating crystals, if applicable
3.2.6　Cooling system(s)
3.2.7　Power supply
3.2.8　Etc..

3.3 How the Laser Components Collectively Produce Laser Radiation.
　　Laser Resonator Types (Stable/Unstable) and Expected Transverse Mode(s) output.
3.3.1　Explain how the resonant cavity architecture dictates the TEM mode output.

3.4 Laser Resonator Fabrication Procedure,

This is the most important part of this procedure so exhaustively cover each step needed for

1. Laser System Alignment Set-up

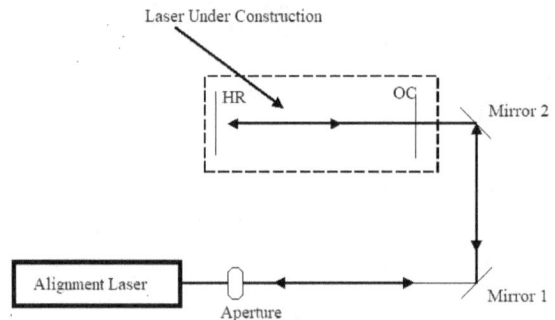

Figure 3.1 Typical laser alignment set-up.

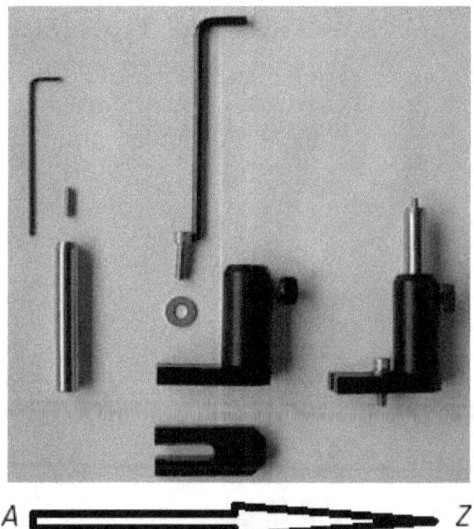

Figure 3.2 Visual illustration of post holder assembly from A to Z

2. Components assembling
3. Support systems testing and installations
4. Walking the alignment beam
5. Getting the first lase and precautions

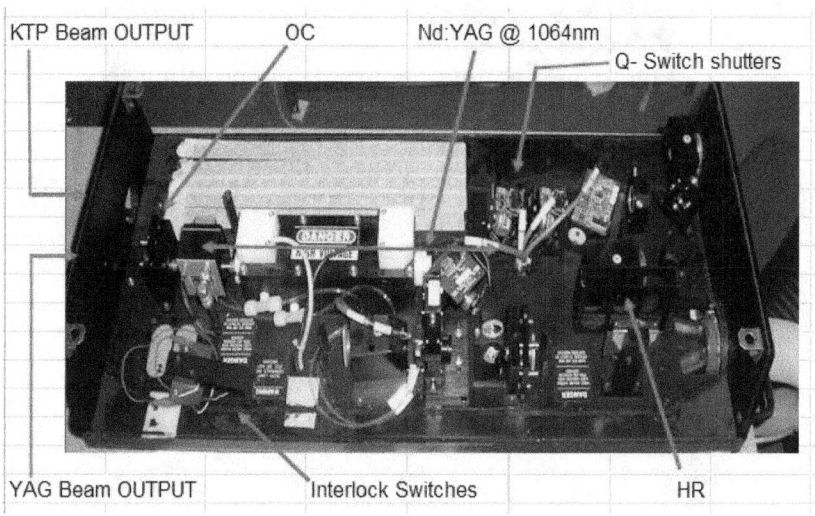

Figure 3.3 Visual overview of the fabricated laser system

    ii. Output power optimization
   iii. Installing and testing the feedback and control system
   iv. Performing output performance tests

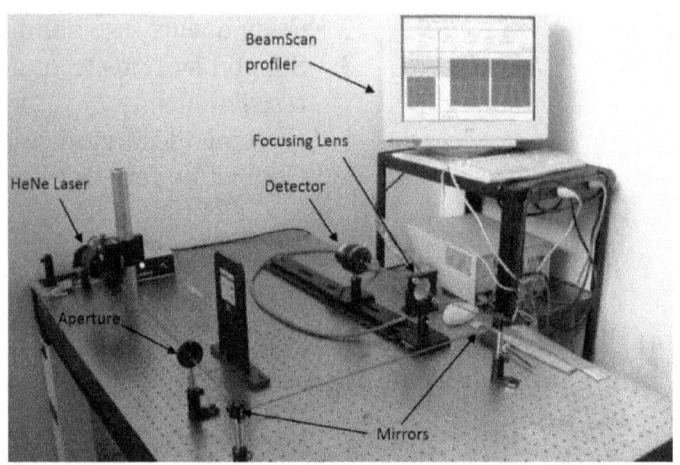

Figure 3.4 Picture of a laser beam profiling set-up

4. Format for Standard Operating Procedures (SOPs)

Getting Started with _____ Laser/Analyzer Procedure
(Do a for each laser and or analyzer operated in this course)

Cover Page/ Example – See Section

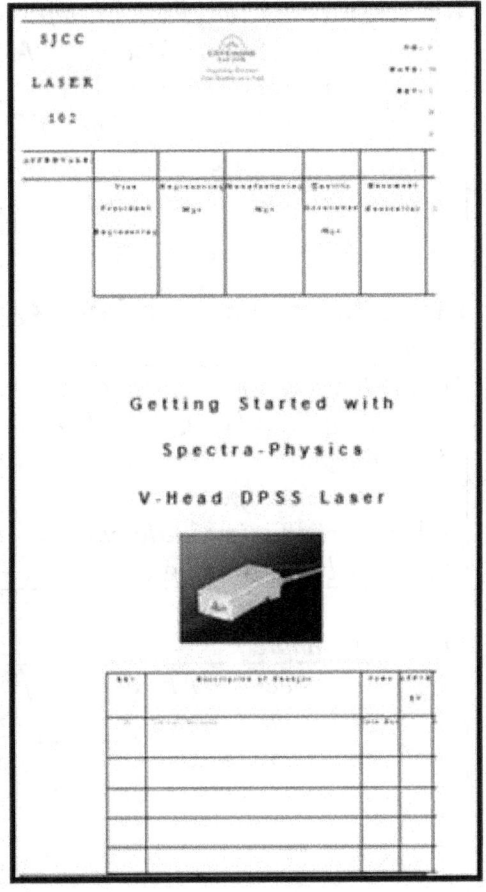

Figure 4.1. Illustration of a test procedure cover page.

1. **Preliminaries** – See Section 1.1

2. **System Overview:** *Example*

   > *E1. ZZZ laser is is a Vanadate-based laser. It is comprised of three main modules: the laser head, controller and Model QQQ power supply*

3. **Device Information,** *laser example*

   > a. *Power Supply Model: V20-10XX*
   > b. *Laser Head Model XXX*
   > c. *Laser Controller Model TTT*
   > d. *Pump source: 808 nm Diode Laser*
   > e. *SHG output wavelength: 532nm exc..*

4. **Device Performance and Safety Controllers**
   Example(s)

   > **E1. A key switch**— *limits access to the laser and prevents it from being turned on for safety reasons. The laser must only operate when the "key" is present and in the "on" position.*
   > **E2. Laser emission indicator**— *when on it means that the laser beam is being emitting.*

5. **Set-up:** *Example*

   > 1. *Connect laser head to the power supply using a BNC cable*

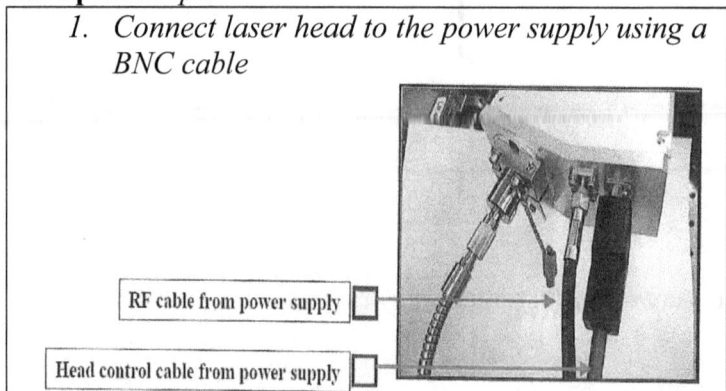

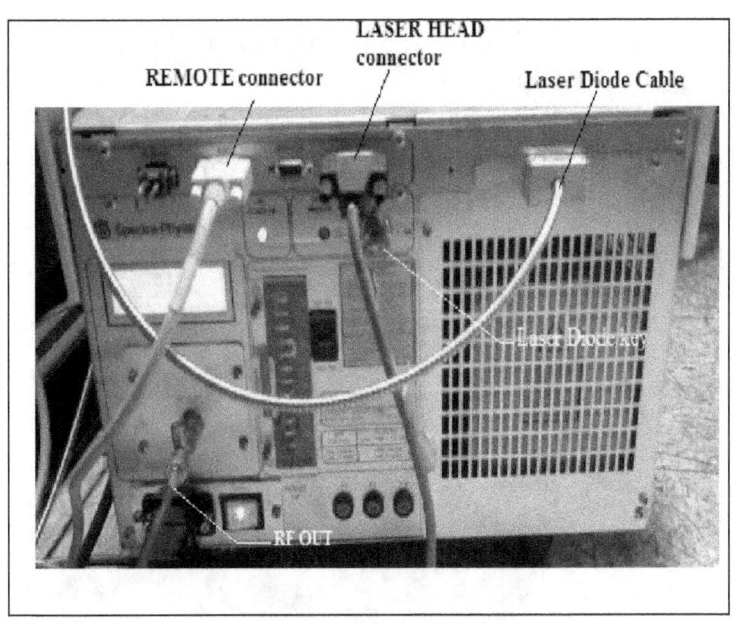

6. Step-by-step Operation Instructions

From start-up, to shut down.

Example
1. Verify power supply is connector.
2. Turn on computer, click on DSU program
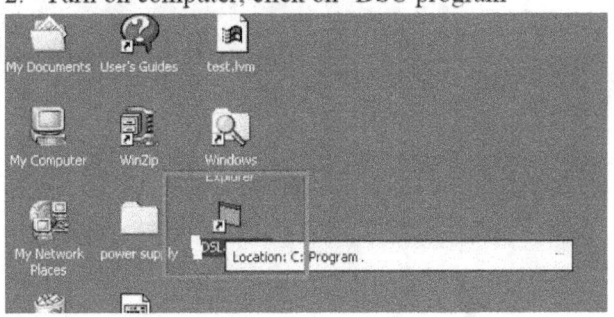
3. Select Controller Interface
Etc..

.

.

10. Fasten bolt to 2 torques

12. Raise the temperature to 30 ^{0}C.
Etc..

If procedure is for laser operation read through section 4.1.

Insert pictures and drawings wherever possible to support your procedure instructions.

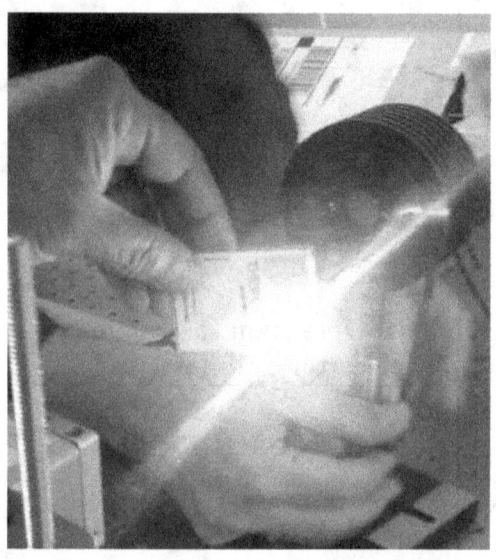

4.1 Laser System Start-up and Shutdown

- There are two main startup sequences namely Cold Start and Warm Start.
  - Cold Start is when the power supply was turned off, while
  - Warm Start is when system is in a standby mode i.e. the laser is off, but the power supply was left on.
- Make it clear as to whether you are addressing cold or warm start.

Turning on Cold/Warm Start

1. State all things that need to be verified before turning on the laser/power supply
2. Chiller/cooling system start-up sequences (if applicable)
3. Power Supply startup sequences (if applicable)
4. Controller start-up sequences (if applicable)
5. Turning on laser sequences
6. State laser warm-up time so that the user knows when to start collecting data
7. Getting the laser in standby mode (optional)

Standby Mode

- It is recommended that you leave the laser in a standby mode when not in use. This means that you always turn off the laser but leave the power supply power switch on. This will keep the SHG crystal (if the laser has one) at the optimal operating temperature, thus reduce warm-up time.

Turning Off Laser
- Laser turn-off sequence
  - If applicable, when to turn off the cooling system. State the time between turning off the laser and turning off the cooling system to avoid damage of laser components from residual heat.

5. Procedure Test Drive

- You must test the procedure you wrote before deploying/publishing it.
- It is recommended that you also have at least one person also test drive the procedures you write even after you have tested them yourself.
    - Your procedures must be accessible to all, including those with disabilities, if not you may be out-of-compliance with some government accessibility laws.
        - https://www.ada.gov/ada_intro.htm
        - https://www.section508.gov/manage/laws-and-policies
- Use the feedback you get to improve the procedure before publishing.

6. Laser Tech Book Catalog, Courses and Certifications

#	ISBN	Title	# of Copies/ Courses Needed
1	ISBN-10: 1724808303 ISBN-13: 978-1724808301	Basic Laser Technology: Comprehensive Course Notes and Workbook	
2	ISBN-10: 1725881314 ISBN-13: 978-1725881310	Solid State Lasers and their Common Problems: Comprehensive Courses Notes and Workbook	
3	ISBN-10: 1796522872 ISBN-13: 978-1796522877	Laser Advanced Concepts, and Common Problems and Practical Solutions: Comprehensive Course Notes and Workbook	
4	ISBN-10: 1727678699 ISBN-13: 978 1727678697	Good Lab and Presentation Practices: Lab Notebook with Tips on How to Protect Intellectual Property (IP)	
5	ISBN-10: 1721540539 ISBN-13: 978-1721540532	Good Laser Lab and Manufacturing Practices (GLLMP): Laser Lab Fundamentals and Performance Tests	
6	ISBN-10: 1725880385 ISBN-13: 978-1725880382	Good Laser Lab and Manufacturing Practices (GLLMPs): Laser Fabrication, Factory-level Tuning and Performance Tests	
7	SBN-10: 1796517658 ISBN-13: 978-1796517651	Good Laser Lab and Manufacturing Practices (GLLMPs): Laser Lab Instrumentation and Opto-Electronic Analyzers	
8	ISBN-10: 1541194543 ISBN-13: 978-1541194540	Laser Transactional/Performance Specifications: General Checklist and Self-Test	

How to Purchase the Books

Single copies of the books can be purchased at Amazon but for volume discounts, 10 or more books, contact your local/Follett bookstore.

We are a registered vendor with Follett Bookstores nationwide https://www.bkstr.com/efollettstore/directory so, see if there is any near you. We can also work with any bookstore of your choice. Have the manager contact us and we will ship the books you want directly to them for resale. Note that you can also get **free book copies** if you order a course, please details below.

orders@laserpronet.com

LaserPronet Laser Tech Certifications
Available 24/7/365 @ www.freelaseriqscan.com

1. **Certificate of Achievement 1: Laser Fundamentals and Performance**
   LP 1.1 Laser Optical Principles
   LP 1.2 Laser Fundamentals
   LP 1.3 Laser Performance

2. **Certificate of Achievement 3: Solid State Lasers and their Common Problems**
   LP 3.1 Laser Pumps, Crystals and Amplifiers
   LP 3.2 Solid State Laser Resonators, Q-switching & Harmonics Generation
   LP 3.3 Laser Head/Resonator and Systems Alignment

Each certificate for less than the cost of a night on the town!

- *Without the benefit of our courses just one employee's single blunder could cost your business more than the price of one of our courses or even millions in revenue.*
- ***One course cost is less than the price of a new laser rod!***

laserpronet.com
Empowering the Laser Workforce

"Book a course and empower yourself/your laser tech workforce TODAY!!"

orders@laserpronet.com